Ajith Thomas

An Overview of Particle Physics

GRIN Publishing

Bibliographic information published by the German National Library:

The German National Library lists this publication in the National Bibliography; detailed bibliographic data are available on the Internet at http://dnb.dnb.de .

Imprint:

Print and binding: Books on Demand GmbH, Norderstedt Germany
ISBN: 978-3-656-84086-2

This book at GRIN:

http://www.grin.com/en/e-book/283818/an-overview-of-particle-physics

AN OVERVIEW OF PARTICLE PHYSICS

AJITH THOMAS

CONTENTS

1. INTRODUCTION

One of the prominent research areas, which have gained a lot of attention of the scientific community as well as that of the general public is the study of the basic structure of matter. 'Elementary particle physics' as it is called is a balancing act of theoretical predictions and experimental confirmations.The widespread attention and media coverage on the latest discoveries which are being announced frequently by CERN, the best particle physics study centre in the world, is undoubtedly due to the inherent interest and curiosity of human beings to understand the framework of the world they are in. This book provides an overview of the things that are going on in the micro world, to an audience who may not be specialised in science, but are interested in understanding the study of 'matter' with a little effort. Special emphazis is given to ' particle accelarators' section, which is the practical side of particle physics research.

2. NUCLEAR COMPOSITION

Protons and neutrons are the macro constituents of the atomic nucleus.Nuclei of the same element have the same number of protons, but can have different number of neutrons.The varieties of an element that differ in the numbers of neutrons their nuclei contain are called isotopes. The protons and neutrons are held together in the nuclei by a short range attractive force called strong force.This force is powerful enough to overcome the electric repulsion of the positive charged protons, provided the neutrons are also present to help. The atomic number (Z) of an element is the number of protons its nucleus pocess, and this will be the same as the number of electrons in that atom, since the atom is electrically neutral.The atomic mass refers to the mass of neutral atoms, not of the bare nuclei. Thus the atomic mass always includes the masses of its Z electrons. A mass number (A) is associated with each element, which is the sum of the number of protons and the number of neutrons in the atom of the element. Atomic masses are expressed in a unit called a.m.u (atomic mass unit) .It is found that the volume of a nucleus is directly proportional to the number of nucleons it contains, which is its mass number.When ever the number of nuterons (indicated by N) is very different from number of protons (Z) in a nucleus, it makes the nucleus unstable.

Nuclear Decay

Nuclear decay occurs in an unstable nuclei, till a final stable one is formed.Nuclear decay occurs in three modes viz. alpha decay, beta decay (positive and negative), and electron capture.Nuclear deacy is not at all a straight forward approach.This involves a lot of elementary particles which are therefore part of nuclear structure. Hundreds of elementary particles have been discovered so far.

3. Fudamental Interactions and associated particles

The Fudamental physical forces in the universe are strong force, electromagnetic force weak force and gravitational force. Force results due to interactions among particles.It is postulated to be occuring by the exchange of another particles. The following table gives a rough picture of the fundamental forces, the details will be dealt with later.

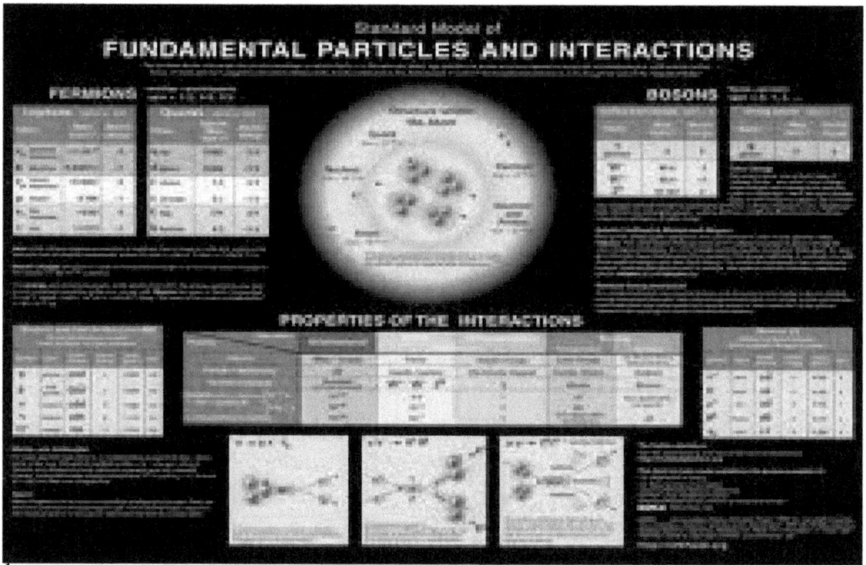

[1]

Name of the Interaction (resulting in a force)	Relative strength	range	particles exchanged.	Role in the universe
Strong	100	10^{-15}	Gluons	holds the quarks together to form nucleons
			Mesons	holds nucleons together to form nuclei
Electromagnetic	1	Infinite	Photons	determines the structures of atoms, molecules solids and liquids.Is the one we feel
Weak	10^{-10}	10^{-18}	W, Z (intermediate bosons)	mediates transformations of quarks and leptons, helps determine compositions of atomic nuclei.
Gravitational	10^{-40}	infinite	gravitons	Assembles matter.

[1] http://www.physi.uni-heidelberg.de/Einrichtungen/FP/anleitungen/F13/gif/cpep_standardmodel.gif

4. Classes of elementary particles

Elementary particles fall into two classes - leptons and hadrons, depending on whether they respond to the strong nuclear force between nucleons.hadrons do respond while leptons do not.Leptons are truly elementary with no hint of internal structures.The electron is a typical example.Hadrons (eg. protons, nuetrons) occupy space and are composed of either two or three quarks, which like leptons are structureless and elementary as present measurements can establish.The strong force that acts between hadrons is the external manifestation of the more basic interactions among the quarks they contain.Hadrons that consist of three quarks, such as proton and neutron are called baryons, and those contain two quarks such as pion are called mesons.There are convincing evidence that quarks exist, but it is not possible to isolate them from hadrons.

5. Leptons

The lepton called electron was the first elementary particle for which a satisfactory theory was developed.It was paul dirac (1928) who obtained a relativistically correct wave equation for a charged particle in an electromagnetic field.An unexpected result of Dirac's theory was its requirement that an electron can have negative as well as positive energies.But the very existence of stable atoms disprove negative energy of electrons. Dirac rescued his theory by postulating that all negative energy states are normally filled. So by a famous principle called Pauli's exclusion principle, it prevents any other electron from dropping into this negative energy 'sea'. But if any electron is given enough energy (by lighting, or when interacted with photons), then this electron can jump out of this negative energy sea, and becomes a positive energy electron. This leaves behind a hole in the negative energy sea, an entity which behaves as a positive charge.The entire process is that a gamma ray (photon) materialises into an electron and positive charged particle. This positive charged particle was later discovered and has been called positron, the anti particle of electron. The other leptons are the muon and tau, and the neutrinos associated with them.

Name of the lepton	Antiparticle name
Electron	Positron
Electron neutrino	Electron antineutrino
Muon	Antimuon
Muon neutrino	Muon antineutrino
Tau	Antitau
Tau neutrino	Tau antineutrino

Spin' property of elementary particles

In quantum mechanics and particle physics, **spin** is a fundamental characteristic property of elementary particles.

All elementary particles of a given kind have the same **spin quantum number**, an important part of the quantum state of a particle., The spin of electrons results in the Pauli's exclusion principle Wolfgang Pauli was the first to propose the concept of spin, but he did not name it. In 1925, Ralph Kronig, George Uhlenbeck, and Samuel Goudsmit suggested a physical interpretation of particles spinning around their own axis. The mathematical theory was worked out in depth by Pauli in 1927. When Paul Dirac derived his relativistic quantum mechanics in 1928, electron spin was an essential part of it.

There are two types of angular momentum in quantum mechanics: Orbital angular momentum, which is a generalization of angular momentum in classical mechanics (**L=r×p**), and spin, which has no analogue in classical mechanics. Since spin is a type of angular momentum, it has the same dimensions: Js in Si units In practice, however, SI units are never used to describe spin: instead, it is written as a multiple of the reduced Planck constant ħ. In natural units, the ħ is omitted, so spin is written as a unit less number. The spin quantum numbers are always unit less numbers by definition.

Neutrino and anti neutrino

A neutrino is one of the elementary particles, a lepton with zero charge, spin ½ and extremely small mass. Neutrinos come in three varieties, each associated with an electron-like lepton: the electron neutrino, the muon neutrino and the tau neutrino.

The need for neutrinos (sticky electron neutrinos) was first pointed out by Wolfgang Pauli, in 1930, to explain the missing energy in beta decay (Also to conserve the linear momentum) . Later in 1956 neutrino were proved to exist, and in 1980's cosmologists started to explore the possibility that neutrinos may make up some of the dark matter in the universe.Lacking charge and mass, and not electromagnetic in nature as the photo is, the neutrino can pass unimpeded through vast amounts of matter.A neutrino would have to pass through over 100 light years of solid iron on the average before interacting.

Distinction between neutrino and anti neutrino

The spin of the neutrino is anticlockwise, when viewed from behind, while the spin of antineutrino is clockwise.Experiments on them showed that, neutrinos and anti neutrinos are distinguishable, having right handed and left handed spins respectively- on all other things they are the same.

6. Hadrons

Unlike leptons, hadrons are subjected to the strong interaction and they have known internal structures.They are of two types mesons and baryons as already mentioned.Mesons consists of a quark and an anti quark, about 140 types are known.Baryons consists of three quarks, about 120 types are known.The lightest baryon is the proton, which is also the only apparantly stable hadron in free space.All baryons other than nucleons decay with mean lives of less than 10^{-9} seconds, in a variety of ways, but the end result is always a proton or neutron.

Resonance particles.

Most hadrons have a measurable time period existence.But some experiments prove that there are also hadrons, which are very short lived, that it is difficult to measure their life time directly.They are called resonance particles.. The lifetime of these particles is on the order of 10^{-23} seconds. Traveling at the speed of light, these particles could only travel about 10^{-15} meters, or about the diameter of a proton, before decaying. Distances of this magnitude cannot be measured in bubble chambers or any other device for detecting subatomic particles. How can we know anything about particles we cannot detect? To understand how we deduce properties of resonance particles, it is first necessary to examine another, more complicated, explanation of their existence. This explanation involves the 'scattering cross-section 'of a particle. When two particles move towards each other and collide, it is possible to say that the collision was caused by the cross-section of the particles. The greater the cross-section of the particles is, the more likely it is that there will be a collision. So, if we have two beams of particles, the amount of scattering that occurs is related to the cross-section of the particles that make up the beams (this is a simplification, but it helps to understand resonances) . We can measure the cross-section of a particle by knowing how much scattering occurs when two beams of particles collide.

If we graph the results of our observations for the cross-section of the particles versus the total energy of the particles, we can see that the graphs have peaks and valleys. This means that the cross-section of the colliding particles changes as a function of the total energy in the collision. Most collisions have several possible outcomes, and each possibility has peak cross-sections at certain energies. When graphs of the different possible results of the same collision are compared, we find that the peak cross-sections occur at the same energies for each possibility. There must be some reason why all the peaks occur at the same energy.

There are two explanations for the peaks, both involving resonances. In one view, the peaks themselves are resonant states or resonances. The resonance is the peak itself, not a particle. Resonances are simply energies at which the cross section of a particle reaches a maximum. In this view, resonances are similar to atomic energy levels, the only difference being that energy levels can be explained by quantum electro-magnetic theory and the need for discussing peak electron levels at certain energies is gone. Elementary particles are not as well understood, so most of the information we have comes from the resonances.

The second explanation says that the peaks are evidence for actual particles that form as intermediate steps in the collision. In this view, the presence of resonance particles adds to the cross-section of the particles in the collision, making the collision more likely. The peaks are interpreted as evidence for the presence of resonance particles, and the different peaks are caused by a large number of distinct particles, which are just as real as other particles, the only difference being a difference in lifetime.

Both explanations, resonant states and resonance particles, have their advantages, and either can be used to find resonance properties. The energy of resonaces is easy to find; it is just the energy at which the cross-section reaches a peak. In the particle theory, the energy is the mass of an intermediate particle through which the reaction takes place. The particle is formed by the collision but almost instantly decays into more stable particles. According to the resonance explanation, the energy is a resonant state of the reaction between the colliding particles, an energy at which the collision is more probable.

Finding the lifetime of a resonance is also fairly uncomplicated. According to the uncertainty principle, dE*dt is greater than or equal to h/2. The mean lifetime is given by t=h/dE. This is the same formula used to find lifetimes of excited nuclear states. dE is the width of the peak at the half maximum. If resonances are particles, then this formula gives their lifetimes. If they are resonant states, then the lifetime is the duration of the resonance, which is harder to understand. Again, the comparison to atomic energy states is useful. The time of the resonance is analogous to the time an electron stays in an excited state.The debate over the nature of resonances will probably last until we have equipment sensitive enough to measure the extremely short distances which resonances would travel if they were particles. Whether particle or resonant state is of little practical importance, since we can measure the properties of resonances whichever explanation we accept.

Elementary particle quantum numbers

The interactions and decays of the hundreds of known elementary particles and resonances form what seems to be a bewildering array.Order can be brought into this situation by assigning certain quantum numbers to each entity and establishing which of these numbers are conserved and which can change in a given process.The baryon number 1 is assigned to all the baryons, and B=-1 to all antibaryons, all other particles have B=0.The lepton number L_e=1 is assigned to electron and the e-neutrino, and L_e= -1 to to their antiparticles, all other particles have L_e=0.The significance of these numbers is that, in every process of what ever kind, the total values of B and L_e remain constant.Another quantum number used is the strangeness number which is conserved in all the processes mediated by the strong and electromagnetic interactions.

7. Quarks

The quark model was independently proposed by physicists Murray Gell-Mann and George Zweig in 1964. The proposal came shortly after Gell-Mann's 1961 formulation of a particle classification system known as the *Eightfold Way.*At the time of the quark theory's inception, the "particle zoo" included, amongst other particles, a multitude of hadrons. Gell-Mann and Zweig posited that they were not elementary particles, but were instead composed of combinations of quarks and antiquarks. Their model involved three flavors of quarks—up, down and strange—to which they ascribed properties such as spin and electric charge. The initial reaction of the physics community to the proposal was mixed. There was particular contention about whether the quark was a physical entity or an abstraction used to explain concepts that were not properly understood at the time.

In less than a year, extensions to the Gell-Mann–Zweig model were proposed. Sheldon Lee Glashow and James Bjorken predicted the existence of a fourth flavor of quark, which they called *charm*. The addition was proposed because it allowed for a better description of the weak interaction (the mechanism that allows quarks to decay), equalized the number of known quarks with the number of known leptons, and implied a mass formula that correctly reproduced the masses of the known mesons.

In 1968, deep inelastic scattering experiments at the Stanford Linear Accelerator Center (SLAC) showed that the proton contained much smaller, point-like objects and was therefore not an elementary particle. This was the most direct experiments which pointed to the reality of existence of quarks. Physicists were reluctant to identify these objects with quarks at the time, instead calling them "partons"—a term coined by Richard Feynman. The objects that were observed at SLAC would later be identified as up and down quarks as the other flavors were discovered.Nevertheless, "parton" remains in use as a collective term for the constituents of hadrons (quarks, antiquarks, andgluons) .

The strange quark's existence was indirectly validated by SLAC's scattering experiments: not only was it a necessary component of Gell-Mann and Zweig's three-quark model, but it provided an explanation for the kaon (K) and pion (π) hadrons discovered in cosmic rays in 1947.

In a 1970 paper, Glashow, John Iliopoulos and Luciano Maiani presented further reasoning for the existence of the as-yet undiscovered charm quark. The number of supposed quark flavors grew to the current six in 1973, when Makoto Kobayashi and Toshihide Maskawa noted that the experimental observation of CP violation. could be explained if there were another pair of quarks. (In particle physics, **CP violation** is a violation of the postulated **CP-symmetry**: the combination of C-symmetry (charge conjugation symmetry) and P-symmetry (parity symmetry) . CP-symmetry states that the laws of physics should be the same if a particle were interchanged with its antiparticle (C symmetry), and then left and right were swapped (P symmetry) . The discovery of CP violation in 1964 in the decays of neutral kaons resulted in the Nobel Prize in Physics in 1980 for its discoverers James Cronin and Val Fitch.It plays an important role both in the attempts of cosmology to explain the dominance of matter over antimatter in the present Universe, and in the study of weak interactions in particle physics.)

Charm quarks were produced almost simultaneously by two teams in November 1974, one at SLAC under Burton Richter, and one at Brookhaven National Laboratory under Samuel Ting. The charm quarks were observed bound with charm antiquarks in mesons. The two parties had assigned the discovered meson two different symbols, J and ψ; thus, it became formally known as the J/ψ meson. The discovery finally convinced the physics community of the quark model's validity.In the following years a number of suggestions appeared for extending the quark model to six quarks. Of these, the 1975 paper by Haim Harari was the first to coin the terms *top* and *bottom* for the additional quarks.In 1977, the bottom quark was observed by a team at Fermilab led by Leon Lederman This was a strong indicator of the top quark's existence: without the top quark, the bottom quark would have been without a partner. However, it was not until 1995 that the top quark was finally observed, also by the CDF and DØ] teams at Fermilab. It had a mass much greater than had been previously expected—almost as great as a gold atom.

Additional properties of Quarks

A serious problem with the idea that baryons are composed of quarks was that the presence of two or three quarks of the same kind in a particular particle violates the exclusion principle. (Since quarks are fermions having half integer spin, they are subjected to this principle) . To get around this problem, it was suggested taht quarks and antiquarks have an additional property of some kind that can be manifested in a total of six different ways, analogous to the manifestation of electric charge in two different ways viz. positive and negative.In the case of quarks, this property came to be known as 'colour', and its three possibilities were called red, green and blue.The antiquark colours are antired, antigreen and antiblue.. To obey the exclusion principle, all three quarks in a baryon should have different colours, and meson consists of quark of one colour and antiquark of opposite colour.The conservation here which should be obeyed is that, when the colours are combined, hadrons and antihadrons should be colourless, just as an atom is electrically neutral, when the positive and negative charges are combined.The quark colour is also responsible for the strong interactions, just as electromagnetic interactions are based on electric charge.Colour force aspects prevent the quarks from existing independly outside hadrons.Since hadrons seem to be composed of quarks, the strong interaction between hadrons should ultimately be traceable to an interaction between quarks.The particle sthat quarks exchange to produce this interaction are called gluons, of which eight have been poatulated..Quark colour is the basis for a theory, which deals with the interaction of quarks, known as quantum chromodynamics.

8. The Standard Model

The **Standard Model** of particle physics is a theory concerning the electromagnetic, weak, and strong nuclear interactions, which mediate the dynamics of the known subatomic particles. Developed throughout the mid to late 20th century, The Standard Model is truly “a tapestry woven by many hands”, sometimes driven forward by new experimental discoveries, sometimes by theoretical advances. It was a collaborative effort in the largest sense, spanning continents and decades. The current formulation was finalized in the mid 1970s upon experimental confirmation of the existence ofquarks. Since then, discoveries of the bottom quark (1977), the top quark (1995), and the tau neutrino (2000) have given further credence to the Standard Model. More recently, (2011-2012) the apparent detection of the Higgsboson completes the set of predicted particles. Because of its success in explaining a wide variety of experimental results, the Standard Model is sometimes regarded as a "theory of almost everything".

The Standard Model falls short of being a complete theory of fundamental interactions because it does not incorporate the full theory of gravitation as described by general relativity, or predict the accelerating expansion of the universe (as possibly described by dark energy) . The theory does not contain any viable dark matter particle that possesses all of the required properties deduced from observational cosmology. Nevertheless, the Standard Model is important to theoretical and experimental article physicists alike. For theorists, the Standard Model is a paradigm of a quantum field theory, which exhibits a wide range of physics including spontaneous symmetry breaking, anomalies, non-perturbative behavior, etc. It is used as a basis for building more exotic models that incorporate hypothetical

particles, extra dimensions, and elaborate symmetries (such as supersymmetry) in an attempt to explain experimental results at variance with the Standard Model, such as the existence of dark matter and neutrino oscillations. In turn, experimenters have incorporated the Standard Model into simulators to help search for new physics beyond the Standard Model

Recently, the Standard Model has found applications in fields besides particle physics, such as astrophysics, cosmology, and nuclear physics.

The first step towards the Standard Model was Sheldon Glashow's discovery in 1960 of a way to combine the electromagnetic and weak interactions In 1967 Steven Weinberg and Abdus Salam incorporated the Higgs mechanism into Glashow's electroweak theory, giving it its modern form.

The Higgs mechanism is believed to give rise to the masses of all the elementary particles in the Standard Model. This includes the masses of the W and Z bosons, and the masses of the fermions, i.e. the quarks and leptons.

After the neutral weak currents caused by Z boson boson exchange were discovered at CERN in 1973, the electroweak theory became widely accepted and Glashow, Salam, and Weinberg shared the 1979 Nobel Prize in Physics for discovering it. The W and Z boson were discovered experimentally in 1981, and their masses were found to be as the Standard Model predicted.

The theory of the strong interaction, to which many contributed, acquired its modern form around 1973–74, when experiments confirmed that the hadrons were composed of fractionally charged quarks.

The Higgs boson

The Higgs particle is a massive scalar elementary particle theorized by Robert Brout, François Englert,, Peter Higgs, Gerald Guralnik, C. R. Hagen, and Tom Kibble in 1964 (see 1964 PRL symmetry breaking papers) and is a key building block in the Standard Model.It has no intrinsic spin, and for that reason is classified as a boson (like the gauge bosons, which haveinteger spin) .

The Higgs boson plays a unique role in the Standard Model, by explaining why the other elementary particles, except the photon and gluon, are massive. In particular, the Higgs boson would explain why the photon has no mass, while the W and Z bosons are very heavy. Elementary particle masses, and the differences between electromagnetism (mediated by the photon) and the weak force (mediated by the W and Z bosons), are critical to many aspects of the structure of microscopic (and hence macroscopic) matter. In electroweak theory, the Higgs boson generates the masses of the leptons (electron, muon, and tau) and quarks. As the Higgs boson is massive, it must interact with itself.

Because the Higgs boson is a very massive particle and also decays almost immediately when created, only a very high energy particle accelerator can observe and record it. Experiments to confirm and determine the nature of the Higgs boson using the Large Hadron Collider (LHC) at CERN began in early 2010, and were performed at Fermilab's Tevatronn until its closure in late 2011. Mathematical consistency of the Standard Model requires that any mechanism capable of generating the masses of elementary particles become visible at energies above 1.4 TeV therefore, the LHC (designed to collide two 7 to 8 TeV proton beams) was built to answer the question of whether the Higgs boson actually exists.

On 4 July 2012, the two main experiments at the LHC (ATLAS and CMS) both reported independently that they found a new particle with a mass of about 125 GeV/c^2 (about 133 proton masses, on the order of 10^{-25} kg), which is "consistent with the

Higgs boson." Although it has several properties similar to the predicted "simplest" Higgs, they acknowledged that further work would be needed to conclude that it is indeed the Higgs boson, and exactly which version of the Standard Model Higgs it best supported if confirmed.

9. Dark Matter

In astronomy and cosmology, **dark matter** is a type of matter hypothesized to account for a large part of the total mass in the universe. Dark matter cannot be seen directly with telescopes; evidently it neither emits nor absorbs light or other electromagnetic radiation at any significant level. Instead, its existence and properties are inferred from its gravitational effects on visible matter, radiation, and the large scale structure of the universe. Dark matter is estimated to constitute 84% of the matter in the universe and 23% of the mass-energy.

Dark matter came to the attention of astrophysicists due to discrepancies between the mass of large astronomical objects determined from their gravitational effects, and the mass calculated from the "luminous matter" they contain; such as stars, gas and dust. It was first postulated by Jan Oort in 1932 to account for the orbital velocities of stars in the Milky Way and Fritz Zwicky in 1933 to account for evidence of "missing mass" in the orbital velocities of galaxies in clusters. Subsequently, other observations have indicated the presence of dark matter in the universe, including the rotational speeds of galaxies, gravitational lensing of background objects by galaxy clusters such as the Bullet Cluster, and the temperature distribution of hot gas in galaxies and clusters of galaxies. According to consensus among cosmologists, dark matter is composed primarily of a new, not yet characterized, type of subatomic particle The search for this particle, by a variety of means, is one of the major efforts in particle physics today.[1]

Although the existence of dark matter is generally accepted by the mainstream scientific community, several alternative theories have been proposed to try to explain the anomalies for which dark matter is intended to account.

Dark matter's existence is inferred from gravitational effects on visible matter and gravitational lensing of background radiation, and was originally hypothesized to account for discrepancies between calculations of the mass of galaxies, clusters of galaxies and the entire universe made through dynamical andgeneral relativistic means, and calculations based on the mass of the visible "luminous" matter these objects contain: stars and the gas and dust of the interstellar and intergalacticvmedium.

The most widely accepted explanation for these phenomena is that dark matter exists and that it is most likely composed of heavy particles that interact only through gravity and possibly the weak force; however, alternate explanations have been proposed, and there is not yet sufficient experimental evidence to determine which is correct. Many experiments to detect proposed dark matter particles through non-gravitational means are underway.

According to observations of structures larger than solar systems, as well as Big Bang cosmology interpreted under the Friedmann equations and the FLRW metric, dark matter accounts for 23% of the mass-energy content of the observable universe. In comparison, ordinary matter accounts for only 4.6% of the mass-energy content of the observable universe, with the remainder being attributable to dark energy From these figures, dark matter constitutes 83%, (23/ (23+4.6)), of the matter in the universe, whereas ordinary matter makes up only 17%.

Dark matter plays a central role in state-of-the-art modeling of structure formation and galaxy evolution, and has measurable effects on the anisotropies observed in the cosmic microwave background. All these lines of evidence suggest that galaxies, clusters of galaxies, and the universe as a whole contain far more matter than that which interacts with electromagnetic radiation.

As important as dark matter is thought to be in the cosmos, direct evidence of its existence and a concrete understanding of its nature have remained elusive. Though the theory of dark matter remains the most widely accepted theory to explain the anomalies in observed galactic rotation, some alternative theoretical approaches have been developed which broadly fall into the categories of modified gravitational laws and quantum gravitational laws.

10.PARTICLE ACCELARATORS

A **particle accelerator** is a device that uses electromagnetic fields to propel charged particles to high speeds and to contain them in well-defined beams.

There are two basic classes of accelerators, known as electrostatic and oscillating field accelerators. Electrostatic accelerators use static electric fields to accelerate particles. A small-scale example of this class is the cathode ray tube in an ordinary old television set. Other examples are the Cockcroft–Walton generator and the Van de Graaf generator. The achievable kinetic energy for particles in these devices is limited by electrical breakdown. Oscillating field accelerators, on the other hand, use radio frequency electromagnetic fields and circumvent the breakdown problem. This class, which development started in the 1920s, is the basis for all modern accelerator concepts and large-scale facilities. Rolf Widerøe, Gustav Ising, Leó Szilárd, Donald Kerst and Ernest Lawrence are considered as pioneers of this field, conceiving and building the first operational linear particle accelerator, the betatron, and the cyclotron.

Alongside their best known use in particle physics as colliders (e.g. LHC, RHIC, Tevatron), particle accelerators are used in a large variety of applications, including particle therapy for oncological purposes, and as synchrotron light sources for fields such as condensed matter physics.

Because colliders can give evidence on the structure of the subatomic world, accelerators were commonly referred to as **atom smashers**in the 20th century. Despite the fact that most accelerators (but ion facilities) actually propel subatomic particles, the term persists in popular usage when referring to particle accelerators in general.

Beams of high-energy particles are useful for both fundamental and applied research in the sciences, and also in many technical and industrial fields unrelated to fundamental research. It has been estimated that there are approximately 26, 000 accelerators worldwide. Of these, only about 1% are research machines with energies above 1 GeV, while about 44% are for radiotherapy, 41% for ion implantation, 9% for industrial processing and research, and 4% for biomedical and other low-energy research.

The largest particle accelerators with the highest particle energies, such as the RHIC, the Large Hadron Collider (LHC) at CERN (which came on-line in mid-November 2009 and the Tevatron, are used for experimental particle physics

For the most basic inquiries into the dynamics and structure of matter, space, and time, physicists seek the simplest kinds of interactions at the highest possible energies. These typically entail particle energies of many GeV and the interactions of the simplest kinds of particles: leptons (e.g. electrons and positrons) and quarks for the matter, or photons and gluons for the field quanta. Since isolated quarks are experimentally unavailable due to color confinement, the simplest available experiments involve the interactions of, first, leptons with each other, and second, of leptons with nucleons, which are composed of quarks and gluons. To study the collisions of quarks with each other, scientists resort to collisions of nucleons, which at high energy may be usefully considered as essentially 2-body interactions of the quarks and gluons of which they are composed. Thus elementary particle physicists tend to use machines creating beams of electrons, positrons, protons, and anti-protons, interacting with each other or with the simplest nuclei (e.g., hydrogen ordeuterium) at the highest possible energies, generally hundreds of GeV or more. Nuclear physicists and cosmologists may use beams of bare atomic nuclei, stripped of electrons, to investigate the structure, interactions, and properties of the nuclei themselves, and ofcondensed matter at extremely high temperatures and densities, such as might have occurred in the first moments of the Big Bang. These investigations often involve collisions of heavy nuclei – of atoms like iron or gold – at energies of several GeV per nucleon.

Particle accelerators can also produce proton beams, which can produce proton-rich medical or research isotopes as opposed to the neutron-rich ones made in fission reactors; however, recent work has shown how to make ^{99}Mo, usually made in reactors, by accelerating isotopes of hydrogen although this method still requires a reactor to produce tritium. An example of this type of machine is LANSCE at Los Alamos.

Synchrotron radiation

Besides being of fundamental interest, high energy electrons may be coaxed into emitting extremely bright and coherent beams of high energy photons via synchrotron radiation which have numerous uses in the study of atomic structure, chemistry, condensed matter physics, biology, and technology. Examples include the ESRF, which has recently been used to extract detailed 3-dimensional images of insects trapped in amber. Thus there is a great demand for electron accelerators of moderate (GeV) energy and high intensity.

Low-energy machines and particle therapy

Everyday examples of particle accelerators are cathode ray tubes found in television sets and X-ray generators. These low-energy accelerators use a single pair of electrodes with a DC voltage of a few thousand volts between them. In an X-ray generator, the target itself is one of the electrodes. A low-energy particle accelerator called an ion implanter is used in the manufacture of integrated circuits.

At lower energies, beams of accelerated nuclei are also used in medicine, as for the treatment of cancer.

DC accelerator types capable of accelerating particles to speeds sufficient to cause nuclear reactions are Cockcroft-Walton generators or voltage multipliers, which convert AC to high voltage DC, or Van de Graaff generators that use static electricity carried by belts.

Historically, the first accelerators used simple technology of a single static high voltage to accelerate charged particles. While this method is still extremely popular today, with the electrostatic accelerators greatly out-numbering any other type, they are more suited to lower energy studies owing to the practical voltage limit of about 30 MV (when the accelerator is placed in a gas with highdielectric strength, such as sulfur hexafluoride, allowing the high voltage) . The same high voltage can be used twice in a *tandem accelerator* if the charge of the particles can be reversed while they are inside the terminal; this is possible with the acceleration of atomic nuclei by first adding an extra electron or forming an anionic (negatively charged) chemical compound, and then putting the beam through a thin foil to strip off electrons inside the high voltage conducting terminal, making a beam of positive charge.

Although electrostatic accelerators accelerate particles along a straight line, the term linear accelerator is more often associated with accelerators that use oscillating rather than static electric fields. Thus, many accelerators arranged in a straight line are not termed "linear accelerators" but rather "electrostatic accelerators" to differentiate the two cases.

Linear particle accelerators

In a linear accelerator (linac), particles are accelerated in a straight line with a target of interest at one end. They are often used to provide an initial low-energy kick to particles before they are injected into circular accelerators. The longest linac in the world is the Stanford Linear Accelerator, SLAC, which is 3 km (1.9 mi) long. SLAC is an electron-positron collider.

Linear high-energy accelerators use a linear array of plates (or drift tubes) to which an alternating high-energy field is applied.

As the particles approach a plate they are accelerated towards it by an opposite polarity charge applied to the plate. As they pass through a hole in the plate, the polarity is switched so that the plate now repels them and they are now accelerated by it towards the next plate. Normally a stream of "bunches" of particles are accelerated, so a carefully controlled AC voltage is applied to each plate to continuously repeat this process for each bunch.

As the particles approach the speed of light the switching rate of the electric fields becomes so high that they operate at radio frequencies, and so microwave cavities are used in higher energy machines instead of simple plates.

Linear accelerators are also widely used in medicine, for radiotherapy and radiosurgery. Medical grade LINACs accelerate electrons using a klystron and a complex bending magnet arrangement which produces a beam of 6-30 million electron-volt (MeV) energy. The electrons can be used directly or they can be collided with a target to produce a beam of X-rays. The reliability, flexibility and accuracy of the radiation beam produced has largely supplanted the older use of Cobalt-60 therapy as a treatment tool.

Circular or cyclic accelerators

In the circular accelerator, particles move in a circle until they reach sufficient energy. The particle track is typically bent into a circle using electromagnets. The advantage of circular accelerators over linear accelerators (*linacs*) is that the ring topology allows continuous acceleration, as the particle can transit indefinitely. Another advantage is that a circular accelerator is smaller than a linear accelerator of comparable power (i.e. a linac would have to be extremely long to have the equivalent power of a circular accelerator) .

Depending on the energy and the particle being accelerated, circular accelerators suffer a disadvantage in that the particles emit synchrotron radiation. When any charged particle is accelerated, it emits electromagnetic radiation and secondary emissions. As a particle traveling in a circle is always accelerating towards the center of the circle, it continuously radiates towards the tangent of the circle. This radiation is called synchrotron light and depends highly on the mass of the accelerating particle. For this reason, many high energy electron accelerators are linacs. Certain accelerators (synchrotrons) are however built specially for producing synchrotron light (X-rays) .

Since the special theory of relativity requires that matter always travels slower than the speed of light in a vacuum, in high-energy accelerators, as the energy increases the particle speed approaches the speed of light as a limit, but never attains it. Therefore particle physicists do not generally think in terms of speed, but rather in terms of a particle's energy ormomentum, usually measured in electron volts (eV) . An important principle for circular accelerators, and particle beams in general, is that the curvature of the particle trajectory is proportional to the particle charge and to the magnetic field, but inversely proportional to the (typically relativistic) momentum

Cyclotrons

The earliest operational circular accelerators were cyclotrons, invented in 1929 by Ernest O. Lawrence at the University of California, Berkeley. Cyclotrons have a single pair of hollow 'D'-shaped plates to accelerate the particles and a single large dipole magnet to bend their path into a circular orbit. It is a characteristic property of charged particles in a uniform and constant magnetic field B that they orbit with a constant period, at a frequency called the cyclotron frequency, so long as their speed is small compared to the speed of light *c*. This means that the accelerating D's of a cyclotron can be driven at a constant frequency by a radio frequency (RF) accelerating power source, as the beam spirals outwards continuously. The particles are injected in the centre of the magnet and are extracted at the outer edge at their maximum energy.

Cyclotrons reach an energy limit because of relativistic effects whereby the particles effectively become more massive, so that their cyclotron frequency drops out of synch with the accelerating RF. Therefore simple cyclotrons can accelerate protons only to an energy of around 15 million electron volts (15 MeV, corresponding to a speed of roughly 10% of *c*), because the protons get out of phase with the driving electric field. If accelerated further, the beam would continue to spiral outward to a larger radius but the particles would no longer gain enough speed to complete the larger circle in step with the accelerating RF. To accommodate relativistic effects the magnetic field needs to be increased to higher radii like it is done in isochronous cyclotrons. An example for an isochronous cyclotron is the PSI Ring cyclotron which is providing protons at the energy of 590 MeV which corresponds to roughly 80% of the speed of light. The advantage of such a cyclotron is the maximum achievable extracted proton current which is currently 2.2 mA. The energy and current correspond to 1.3 MW beam power which is the highest of any accelerator currently existing.

Synchrocyclotrons and isochronous cyclotrons

A classic cyclotron can be modified to increase its energy limit. The historically first approach was the synchrocyclotron, which accelerates the particles in bunches. It uses a constant magnetic field , but reduces the accelerating field's frequency so as to keep the particles in step as they spiral outward, matching their mass-dependent cyclotron resonance frequency. This approach suffers from low average beam intensity due to the bunching, and again from the need for a huge magnet of large radius and constant field over the larger orbit demanded by high energy.

The second approach to the problem of accelerating relativistic particles is the isochronous cyclotron. In such a structure, the accelerating field's frequency (and the cyclotron resonance frequency) is kept constant for all energies by shaping the magnet poles so to increase magnetic field with radius. Thus, all particles get accelerated in isochronous time intervals. Higher energy particles travel a shorter distance in each orbit than they would in a classical cyclotron, thus remaining in phase with the accelerating field. The advantage of the isochronous cyclotron is that it can deliver continuous beams of higher average intensity, which is useful for some applications. The main disadvantages are the size and cost of the large magnet needed, and the difficulty in achieving the high magnetic field values required at the outer edge of the structure.

Synchrocyclotrons have not been built since the isochronous cyclotron was developed.

Betatrons

Another type of circular accelerator, invented in 1940 for accelerating electrons, is the Betatron, a concept which originates ultimately from Norwegian-German scientist Rolf Widerøe. These machines, like synchrotrons, use a donut-shaped ring magnet (see below) with a cyclically increasing B field, but accelerate the particles by induction from the increasing magnetic field, as if they were the secondary winding in a transformer, due to the changing magnetic flux through the orbit.

Achieving constant orbital radius while supplying the proper accelerating electric field requires that the magnetic flux linking the orbit be somewhat independent of the magnetic field on the orbit, bending the particles into a constant radius curve. These machines have in practice been limited by the large radiative losses suffered by the electrons moving at nearly the speed of light in a relatively small radius orbit.

Synchrotrons

To reach still higher energies, with relativistic mass approaching or exceeding the rest mass of the particles (for protons, billions of electron volts or GeV), it is necessary to use a synchrotron. This is an accelerator in which the particles are accelerated in a ring of constant radius. An immediate advantage over cyclotrons is that the magnetic field need only be present over the actual region of the particle orbits, which is very much narrower than the diameter of the ring. (The largest cyclotron built in the US had a 184-inch-diameter (4.7 m) magnet pole, whereas the diameter of the LEP and LHC is nearly 10 km. The aperture of the two beams of the LHC is of the order of a millimeter.)

However, since the particle momentum increases during acceleration, it is necessary to turn up the magnetic field B in proportion to maintain constant curvature of the orbit. In consequence synchrotrons cannot accelerate particles continuously, as cyclotrons can, but must operate cyclically, supplying particles in bunches, which are delivered to a target or an external beam in beam "spills" typically every few seconds.

Since high energy synchrotrons do most of their work on particles that are already traveling at nearly the speed of light *c*, the time to complete one orbit of the ring is nearly constant, as is the frequency of the RF cavity resonators used to drive the acceleration.

Note also a further point about modern synchrotrons: because the beam aperture is small and the magnetic field does not cover the entire area of the particle orbit as it does for a cyclotron, several necessary functions can be separated. Instead of one huge magnet, one has a line of hundreds of bending magnets, enclosing (or enclosed by) vacuum connecting pipes. The design of synchrotrons was revolutionized in the early 1950s with the discovery of the strong focusing concept.[13][14][15] The focusing of the beam is handled independently by specialized quadrupole magnets, while the acceleration itself is accomplished in separate RF sections, rather similar to short linear accelerators. Also, there is no necessity that cyclic machines be circular, but rather the beam pipe may have straight sections between magnets where beams may collide, be cooled, etc. This has developed into an entire separate subject, called "beam physics" or "beam optics".[16]

More complex modern synchrotrons such as the Tevatron, LEP, and LHC may deliver the particle bunches into storage rings of magnets with constant B, where they can continue to orbit for long periods for experimentation or further acceleration. The highest-energy machines such as the Tevatron and LHC are actually accelerator complexes, with a cascade of specialized elements in series, including linear accelerators for initial beam creation, one or more low energy synchrotrons to reach intermediate energy, storage rings where beams can be accumulated or "cooled" (reducing the magnet aperture required and permitting tighter focusing; see beam cooling), and a last large ring for final acceleration and experimentation.

Electron synchrotrons

Circular electron accelerators fell somewhat out of favor for particle physics around the time that SLAC was constructed, because their synchrotron losses were considered economically prohibitive and because their beam intensity was lower than for the unpulsed linear machines. The Cornell Electron Synchrotron, built at low cost in the late 1960s, was the first in a series of high-energy circular electron accelerators built for fundamental particle physics, culminating in the LEP at CERN.

A large number of electron synchrotrons have been built in the past two decades, as part of synchrotron light sources that emit ultraviolet light and X rays; see below.

Storage rings

For some applications, it is useful to store beams of high energy particles for some time (with modern high vacuum technology, up to many hours) without further acceleration. This is especially true for colliding beam accelerators, in which two beams moving in opposite directions are made to collide with each other, with a large gain in effective collision energy. Because relatively few collisions occur at each pass through the intersection point of the two beams, it is customary to first accelerate the beams to the desired energy, and then store them in storage rings, which are essentially synchrotron rings of magnets, with no significant RF power for acceleration.

Synchrotron radiation sources

Some circular accelerators have been built to deliberately generate radiation (called synchrotron light) as X-rays also called synchrotron radiation, for example the Diamond Light Source which has been built at the Rutherford Appleton Laboratory in England or the Advanced Photon Source at Argonne National Laboratory in Illinois, USA. High-energy X-rays are useful for X-ray spectroscopy of proteins or X-ray absorption fine structure (XAFS) for example.

Synchrotron radiation is more powerfully emitted by lighter particles, so these accelerators are invariably electron accelerators. Synchrotron radiation allows for better imaging as researched and developed at SLAC's SPEAR.

FFAG accelerators

FFAG accelerators, in which a very strong radial field gradient, combined with strong focusing, allows the beam to be confined to a narrow ring, are an extension of the isochronous cyclotron idea that is lately under development.[17] They use RF accelerating sections between the magnets, and so are isochronous for relativistic particles like electrons (which achieve essentially the speed of light at only a few MeV), but only over a limited energy range for protons and heavier particles at sub-relativistic energies. Like the isochronous cyclotrons they achieve continuous beam operation, but without the need for a huge dipole bending magnet covering the entire radius of the orbits.

History of particle accelarators.

Lawrence's first cyclotron was a mere 4 inches (100 mm) in diameter. Later he built a machine with a 60 in diameter pole face, and planned one with a 184-inch diameter, which was, however, taken over for World War II-related work connected with uranium isotope separation; after the war it continued in service for research and medicine over many years.

The first large proton synchrotron was the Cosmotron at Brookhaven National Laboratory, which accelerated protons to about 3 GeV. The Bevatron at Berkeley, completed in 1954, was specifically designed to accelerate protons to sufficient energy to create antiprotons, and verify the particle-antiparticle symmetry of nature, then only strongly suspected. TheAlternating Gradient Synchrotron (AGS) at Brookhaven was the first large synchrotron with alternating gradient, "strong focusing" magnets, which greatly reduced the required aperture of the beam, and correspondingly the size and cost of the bending magnets. The Proton Synchrotron, built at CERN, was the first major European particle accelerator and generally similar to the AGS.

The Stanford Linear Accelerator, SLAC, became operational in 1966, accelerating electrons to 30 GeV in a 3 km long waveguide, buried in a tunnel and powered by hundreds of large klystrons. It is still the largest linear accelerator in existence, and has been upgraded with the addition of storage rings and an electron-positron collider facility. It is also an X-ray and UV synchrotron photon source.

The Fermilab Tevatron has a ring with a beam path of 4 miles (6.4 km) . It has received several upgrades, and has functioned as a proton-antiproton collider until it was shut down due to budget cuts on September 30, 2011. The largest circular accelerator ever built was the LEP synchrotron at CERN with a circumference 26.6 kilometers, which was an electron/positron collider. It achieved an energy of 209 GeV before it was dismantled in 2000 so that the underground tunnel could be used for the Large Hadron Collider (LHC) . The LHC is a proton collider, and currently the world's largest and highest-energy accelerator, expected to achieve 7 TeV energy per beam, and currently operating at half that.

The aborted Superconducting Super Collider (SSC) in Texas would have had a circumference of 87 km. Construction was started in 1991, but abandoned in 1993. Very large circular accelerators are invariably built in underground tunnels a few metres wide to minimize the disruption and cost of building such a structure on the surface, and to provide shielding against intense secondary radiations that occur, which are extremely penetrating at high energies.

Current accelerators such as the Spallation Neutron Source, incorporate superconducting cryomodules. The Relativistic Heavy Ion Collider, and Large Hadron Collider also make use of superconducting magnets and RF cavity resonators to accelerate particles.

Targets and detectors

The output of a particle accelerator can generally be directed towards multiple lines of experiments, one at a given time, by means of a deviating electromagnet. This makes it possible to operate multiple experiments without needing to move things around or shutting down the entire accelerator beam. Except for synchrotron radiation sources, the purpose of an accelerator is to generate high-energy particles for interaction with matter.

This is usually a fixed target, such as the phosphor coating on the back of the screen in the case of a television tube; a piece of uranium in an accelerator designed as a neutron source; or a tungsten target for an X-ray generator. In a linac, the target is simply fitted to the end of the accelerator. The particle track in a cyclotron is a spiral outwards from the centre of the circular machine, so the accelerated particles emerge from a fixed point as for a linear accelerator.

For synchrotrons, the situation is more complex. Particles are accelerated to the desired energy. Then, a fast acting dipole magnet is used to switch the particles out of the circular synchrotron tube and towards the target.

A variation commonly used for particle physics research is a collider, also called a *storage ring collider*. Two circular synchrotrons are built in close proximity – usually on top of each other and using the same magnets (which are then of more complicated design to accommodate both beam tubes) . Bunches of particles travel in opposite directions around the two accelerators and collide at intersections between them. This can increase the energy enormously; whereas in a fixed-target experiment the energy available to produce new particles is proportional to the square root of the beam energy, in a collider the available energy is linear.

Higher energies

At present the highest energy accelerators are all circular colliders, but it is likely that limits have been reached in respect of compensating for synchrotron radiation losses for electron accelerators, and the next generation will probably be linear accelerators 10 times the current length. An example of such a next generation electron accelerator is the 40 km long International Linear Collider, due to be constructed between 2015-2020.

As of 2005, it is believed that plasma wakefield acceleration in the form of electron-beam 'afterburners' and standalone laser pulsers will provide dramatic increases in efficiency within two to three decades. In plasma wakefield accelerators, the beam cavity is filled with a plasma (rather than vacuum) . A short pulse of electrons or laser light either constitutes or immediately trails the particles that are being accelerated. The pulse disrupts the plasma, causing the charged particles in the plasma to integrate into and move toward the rear of the bunch of particles that are being accelerated. This process transfers energy to the particle bunch, accelerating it further, and continues as long as the pulse is coherent.[18]

Energy gradients as steep as 200 GeV/m have been achieved over millimeter-scale distances using laser pulsers[19] and gradients approaching 1 GeV/m are being produced on the multi-centimeter-scale with electron-beam systems, in contrast to a limit of about 0.1 GeV/m for radio-frequency acceleration alone. Existing electron accelerators such as SLACcould use electron-beam afterburners to greatly increase the energy of their particle beams, at the cost of beam intensity. Electron systems in general can provide tightly collimated, reliable beams; laser systems may offer more power and compactness. Thus, plasma wakefield accelerators could be used — if technical issues can be resolved — to both increase the maximum energy of the largest accelerators and to bring high energies into university laboratories and medical centres.

11.Future of particle Physics Research

The ultimate goal, that is pursued in several distinct ways, is to find and understand about 'what physics may lie beyond the standard model'. There are several powerful experimental reasons to expect new physics, including dark matter. There are also theoretical hints that this new physics should be found at accessible energy scales. Furthermore, there may be surprises that will give us opportunities to learn about nature.